CONSERVATION 2000

Radiation and Nuclear Energy

Typeset by J&L Composition Ltd, Filey,
North Yorkshire
and printed in Great Britain by
Bath Press Colourbooks, Blantyre

for the publishers
B. T. Batsford Ltd
4 Fitzhardinge Street
London W1H 0AH
ISBN 0 7134 6574 3

A CIP catalogue record for this book is
available from the British Library

Acknowledgements

The author and publishers would like to
thank the following for permission to
reproduce illustrations: British Nuclear
Fuels for pages 39, 45, 46/47 and 49; JET
Joint Undertaking for pages 26/27 and 28;
T. Norton, MAFF for pages 8, 18/19, 20
and 54/55; National Radiological
Protection Board for pages 6/7, 8, 12,
14/15, 22/23, 24, 30/31, 37, 45, 52 and 53;
Science Photo Library for pages 10/11, 16,
18, 31, 33, 34/35, 36, 38/39, 42/43, 44, 56
and 58/59; Frank Spooner Picture
Library for pages 50/51; Stanford
University Medical Center for page 17.

CONSERVATION 2000

Radiation and Nuclear Energy

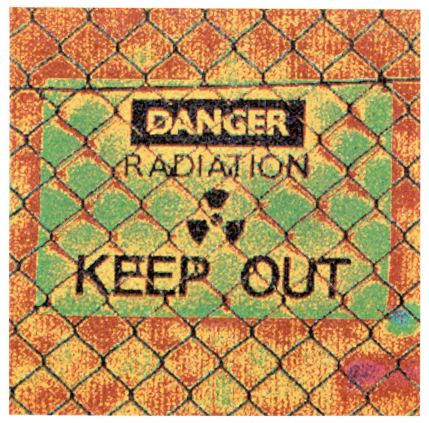

Joy Palmer

B. T. Batsford Ltd London

CONTENTS

INTRODUCTION

Radiation and nuclear power together form a vast and complicated subject. Nuclear scientists spend their working lives researching in this field so that we might have a better understanding of the benefits and problems associated with radioactivity. As well as being complex, it is also a very fascinating subject and one which is a constant source of public interest and concern.

Most people will have heard of the Chernobyl disaster, the most serious event in the history of nuclear power, which killed and injured thousands of people. Sensational occurrences such as this are bound to cause anger and promote controversy over the use of radioactive materials, and, whether we like it or not, radiation in our lives *is* controversial. Yet it is not all bad news. Radiation is part of our natural environment. What is important is to understand what it is, and to develop an informed awareness of its potential dangers as well as its benefits in our everyday lives.

The fear of nuclear war is very real. Many people care deeply about this and argue forcefully that it is wrong to devote more and more resources, including vast amounts of money, to the production and storing of weapons that could destroy our world. This debate will no doubt continue.

This and many other related issues are raised and studied in the following pages. Indeed, the whole subject of radiation and nuclear matters, a source of considerable controversy, is one about which much has already been written elsewhere and in great depth. This book aims to explain, explore and illustrate a range of key issues involved – and to do so without taking sides. It is up to each reader as an individual to consider his or her own point of view in this debate, a debate which must concern each and every one of us. Radiation is inescapable: to a greater or lesser extent, it has some impact upon every living thing inhabiting our Earth.

THE WORLD OF THE ATOM

What is radiation? If you look up the word 'radiate' in your dictionary, you will find that it means 'to spread out rays'. There are many types of radiation, including light and heat. To say that something is radioactive simply means that it is made up of atoms which give off rays or tiny particles of radiation at very high speed.

Radiation is present all around us. All matter is made up of atoms and we live in a radioactive world: the air we breathe, the earth we walk upon, the outer space we look into, and the food and drink we consume are all slightly radioactive. Our television sets emit radiation; we also receive small amounts when undergoing some hospital treatment or diagnosis, from X-rays for example. The small amount of radiation in our natural environment and everyday life is virtually harmless – it does not affect our health in any significant way. However, large doses of radiation are extremely dangerous, as subsequent pages will explain.

Soil samples in proximity to nuclear power stations are tested to see whether they contain an acceptable level of radioactivity.

Everything in our world is made up of atoms. Atoms are so small that they are difficult to describe or imagine. One page of this book, for example, is around one million atoms thick! Because they are so incredibly tiny, it is impossible to separate out the atoms by cutting or dividing objects up. They can best be seen using extremely high-powered instruments called electron microscopes. Every atom is made up of particles even tinier than itself. At the centre is a nucleus made up of protons and neutrons. Around the nucleus are one or more electrons. These electrons move around the nucleus in layers called shells.

Some atoms may be described as 'stable'. Others are 'unstable'. If there is the correct number of neutrons to balance the number of protons present, then the atom is stable. Stable atoms do not change. Most of the world around us is made up of stable atoms.

Unstable atoms, on the other hand, do change and it is these which are termed radioactive.

These atoms have an unstable nucleus because they do not have the correct balance of protons and neutrons. Instability occurs when there are either too many or too few neutrons in the nucleus. When this situation exists, the nucleus produces radiation – that is, tiny particles and rays which are sent out. Radioactive atoms send out various kinds of radiation. These include alpha particles, beta particles and gamma rays.

The electron microscope is used to examine radioactive dust particles from places of work.

Special equipment is needed to monitor radiation in land surveys. Often this is transported by helicopter from one site to another.

An alpha particle is made up of two protons and two neutrons. Streams of these protons and neutrons can only travel a few centimetres in the air and are not very strong. They cannot penetrate paper or skin. Beta particles are electrons that are sent out by atoms. They are more penetrating than alpha particles and may travel a few metres. They will pass through skin, but will be stopped by a thin sheet of glass or metal. Gamma rays are not particles but invisible rays, similar to rays of light, which are very penetrating compared to the others. They will pass right through a human body but may be stopped by thick shields of lead or concrete.

A final form of radiation which should be mentioned is neutron radiation. This is found inside the reactors at nuclear power stations. Neutrons, like gamma rays, are extremely penetrating and thick concrete or water is needed to halt them.

All these forms of radiation which arise as a result of discharges from unstable atoms are known as ionizing radiation.

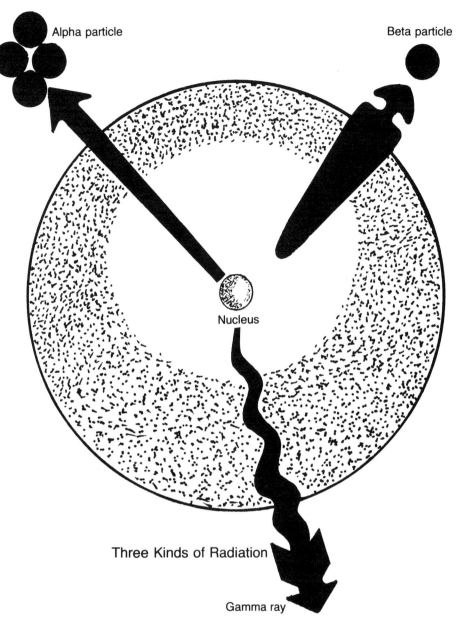

Three Kinds of Radiation

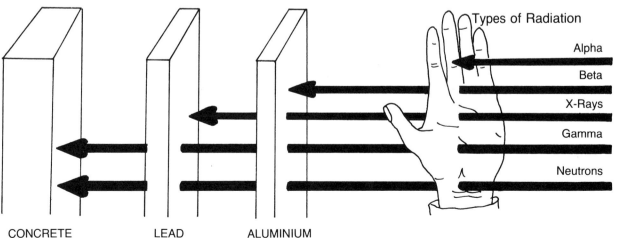

RADIATION ALL AROUND US

Radiation occurs naturally and is all around us, at home and out-doors. Where you live is likely to have an influence on the amount of natural radiation that your body receives: some types of soil are richer in radioactivity than others. There are other sources of radiation too, including cosmic rays from outer space, fall-out

How much radiation?

The amount of radiation absorbed in living matter is called the *dose*.

The absorbed dose is measured with units called *grays*.

Measuring a dose however, is not as straightforward as this. A dose of alpha radiation is about 20 times more dangerous than the same amount of beta or gamma radiation.

A measured dose therefore, is adjusted to take this into account. It is known as the *dose equivalent* and is measured in units called *sieverts* (Sv)

1 Sv = 1,000 millisieverts
1 Sv = 1,000,000 microsieverts.

from nuclear weapons tests, and radiation from medical uses as well as from air travel, luminous watches and television sets. There is also the radiation which occurs naturally in our own bodies and from eating, drinking and breathing. Many people are concerned about the dangers of nuclear power, yet there is no doubt that on a day-to-day basis, most of the radiation we receive comes from our natural environment.

One form in which this background radiation affects us at home is as radon, a gas which has no colour or smell. Radon in the atmosphere is not harmful unless inhaled in exceptionally large quantities. Nevertheless, some scientists are very concerned about the possible dangers of radon gas, which can be a cause of lung cancer. It is thought that radon is the likely cause of lung cancer in people who have never smoked and yet still develop the disease. In 1988 a report in the United States warned that in many homes radon levels were comparable to smoking ten cigarettes a day.

The Environmental Protection Agency estimates that exposure to radon causes between 5,000 and 20,000 of the 136,000 fatal lung cancers suffered each year by US citizens. Some of the highest radon levels in the USA have been found in an area rich in uranium which covers parts of Pennsylvania, New Jersey and New York. Yet other high readings have been found in areas with very little uranium in the soil.

The Environmental Protection Agency is developing a system which can remove radon from the foundations of a building before it can enter the habitable rooms.

In Britain, the Government Department of the Environment issued a leaflet on radon in 1987. Many substances, including stones, bricks, glass, cement and ceramics, may release small quantities of radon into the atomsphere. Some areas of Britain are more exposed to this hazard than others: in Cornwall, Devon and Glasgow, for example, radon levels are higher than in other places because many local houses are built from local granite which bears significant amounts of radon. It is important that those responsible for building homes are aware of this problem, and that careful attention is given to both building materials and to good ventilation.

Radiation is, of course, invisible. We cannot see the rays and particles travelling through the atmosphere, yet scientists are able to detect and to measure them. Several instruments can be used for detecting radiation, so that levels can be checked and

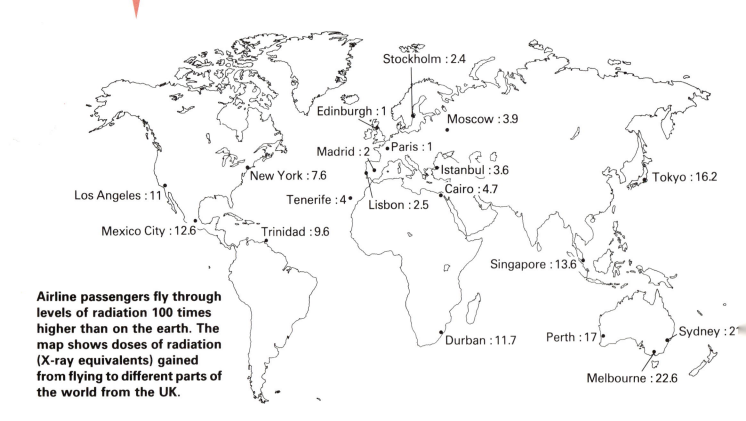

Airline passengers fly through levels of radiation 100 times higher than on the earth. The map shows doses of radiation (X-ray equivalents) gained from flying to different parts of the world from the UK.

Stockholm : 2.4
Edinburgh : 1
Moscow : 3.9
Madrid : 2
Paris : 1
New York : 7.6
Istanbul : 3.6
Los Angeles : 11
Tenerife : 4
Cairo : 4.7
Lisbon : 2.5
Tokyo : 16.2
Mexico City : 12.6
Trinidad : 9.6
Singapore : 13.6
Durban : 11.7
Perth : 17
Sydney : 2?
Melbourne : 22.6

Here a Boeing 737–200 aircraft is being sent up for engine radiography. Radiation is used as a source for the detection of metal fatigue in aircraft. This forms part of regular maintenance.

radioactive substances can be discovered. One useful machine is a geiger counter which can count the number of radioactive particles present. Photographic film is affected by rays and particles, and turns dark if it has been exposed to light. Large monitoring machines are used to measure the amount of radioactivity in a human body.

There is much debate about the dangers of radiation in our everyday lives. Radiation is basically harmful to living things. The rays and particles affect the atoms in living cells, and even relatively small doses of radiation can cause serious disease and death. Because we cannot see it and it can be very dangerous, radiation is rather a mysterious and unnerving subject. Yet while accidents and horrendous catastrophes can and do occur, it is perhaps important to stress that, generally speaking, the background radiation to which we are continually exposed is actually harmless, since it is present in such small quantities. Furthermore, there are many good and positive uses to which radiation may be put.

And what of the effects?

There is no known tolerance level for ionizing radiation. In other words, any dose will cause some harm. The extent of the damage depends on the dose received by an individual.

Scientists are agreed that likely harm to the human body caused by various doses is as follows:

Dose	Effect
10 sieverts	death: in a matter of hours or days
5 to 10 sieverts	death in one or two weeks
3 to 5 sieverts	death to half of the people exposed to this level within two months
1.5 to 2.5 sieverts	severe sickness, skin burns, possible death of foetus in pregnant women, possible permanent health damage
0.5 to 1.5 sieverts	less severe burns and sickness, growth of tumours in body, premature ageing, genetic damage to children of affected people
0.1 to 0.5 sieverts	possible sickness, long term premature ageing, genetic damage, possible risk of tumour growth
less than 0.1 sieverts	long term premature ageing, possible genetic damage and risk of tumours

PUTTING THIS IN THE CONTEXT OF OUR LIVES

One X-ray in certain 'old' machinery (pre 1970) = a dose of 0.03 Sv.

One year's 'average' dose from background radiation = a dose of 0.001 Sv.

One modern X-ray = a dose of 0.0001 Sv.

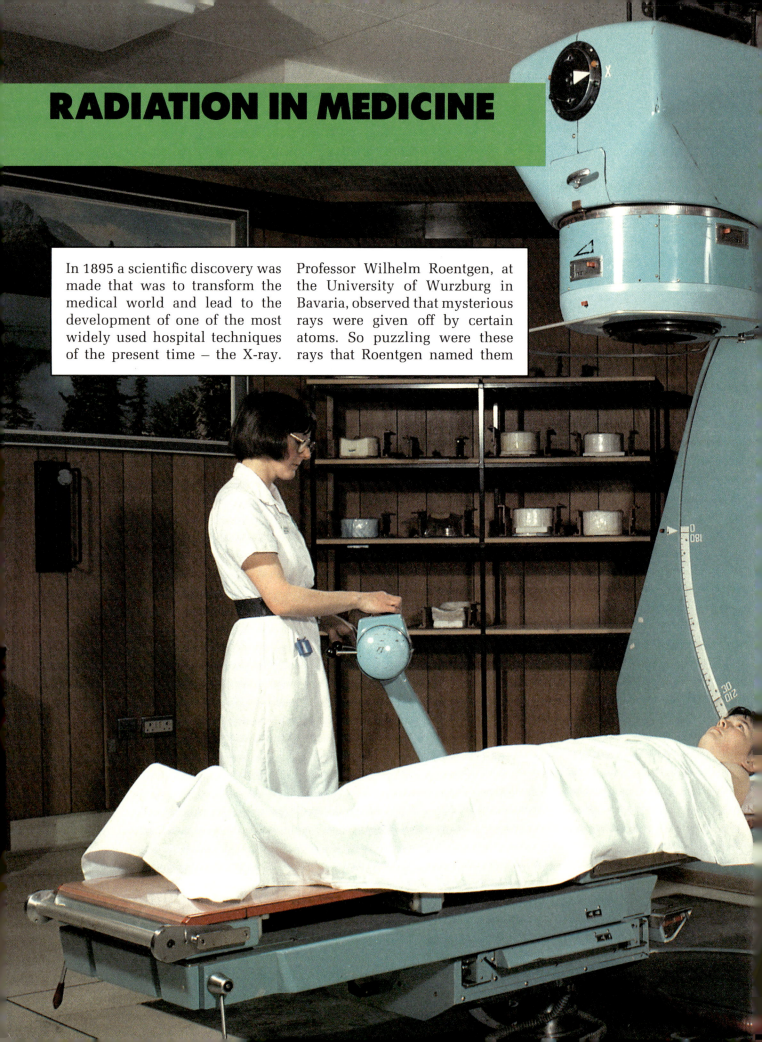

RADIATION IN MEDICINE

In 1895 a scientific discovery was made that was to transform the medical world and lead to the development of one of the most widely used hospital techniques of the present time – the X-ray. Professor Wilhelm Roentgen, at the University of Wurzburg in Bavaria, observed that mysterious rays were given off by certain atoms. So puzzling were these rays that Roentgen named them

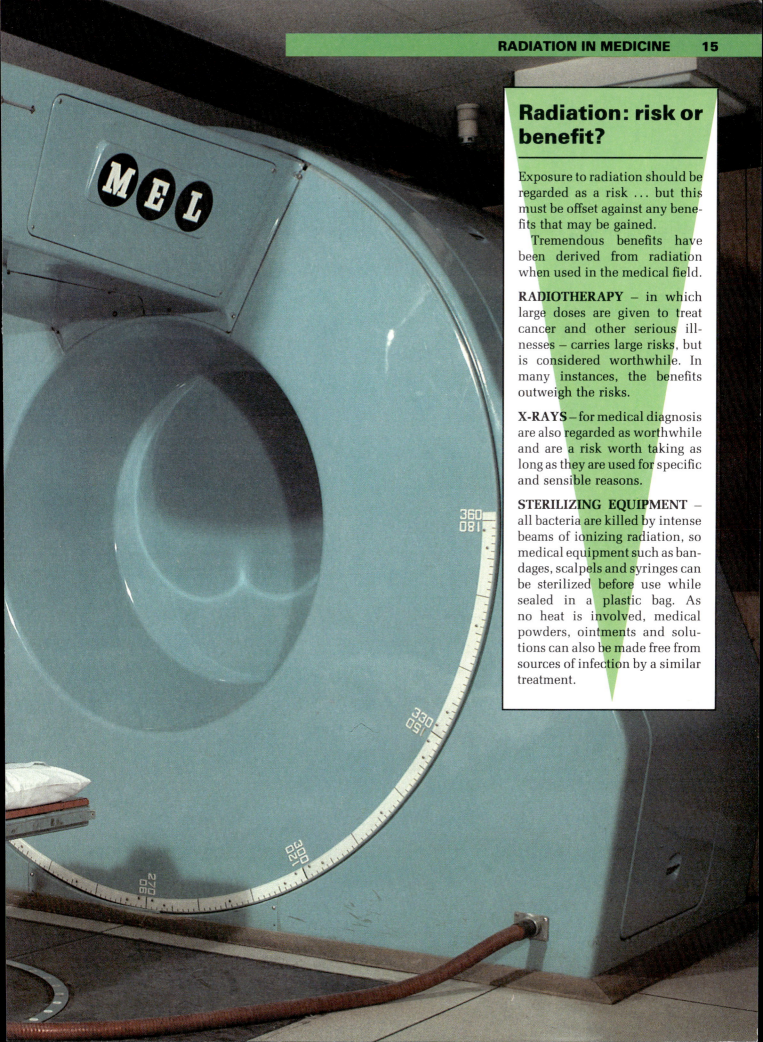

Radiation: risk or benefit?

Exposure to radiation should be regarded as a risk ... but this must be offset against any benefits that may be gained.

Tremendous benefits have been derived from radiation when used in the medical field.

RADIOTHERAPY – in which large doses are given to treat cancer and other serious illnesses – carries large risks, but is considered worthwhile. In many instances, the benefits outweigh the risks.

X-RAYS – for medical diagnosis are also regarded as worthwhile and are a risk worth taking as long as they are used for specific and sensible reasons.

STERILIZING EQUIPMENT – all bacteria are killed by intense beams of ionizing radiation, so medical equipment such as bandages, scalpels and syringes can be sterilized before use while sealed in a plastic bag. As no heat is involved, medical powders, ointments and solutions can also be made free from sources of infection by a similar treatment.

'X-rays', since the symbol 'X' was often used by mathematicians to describe an unknown quantity.

Other scientists were fascinated by Roentgen's work, especially when he demonstrated that X-rays could be passed through the skin of a human body and show the pattern of bones inside. He devised experiments which showed that the invisible radiation would pass through paper, copper and other materials but not through certain metals, glass and human bones. Roentgen was able to take X-ray pictures by placing various objects between the source of the rays and a photographic plate. One of his best-known early pictures was that of his wife's hand, showing only the bones and wedding ring through which the rays did not pass.

Without doubt this was a sensational discovery. News of it soon spread through the scientific world and for a time the radiation in question was known as 'Roentgen Rays'. Other scientists, including Thomas Alva Edison who was famous for his work on electric light and power, followed Roentgen's discovery with their own work on X-rays.

The medical profession greeted Roentgen's discovery with great excitement, seeing the enormous potential for diagnosing illness associated with the human skeleton. Within a year of X-rays being discovered, more than a thousand articles and fifty books had been published on the subject. Soon doctors used X-rays to try to cure diseases as well as to diagnose them, and met with some success. No one knew why, but apparently some patients suffering from cancer and tuberculosis were mysteriously cured as a result of exposure to X-rays. Unfortunately, the cure was sometimes as painful as the disease and it became obvious that X-rays could cause serious medical problems as well as diagnosing and curing them.

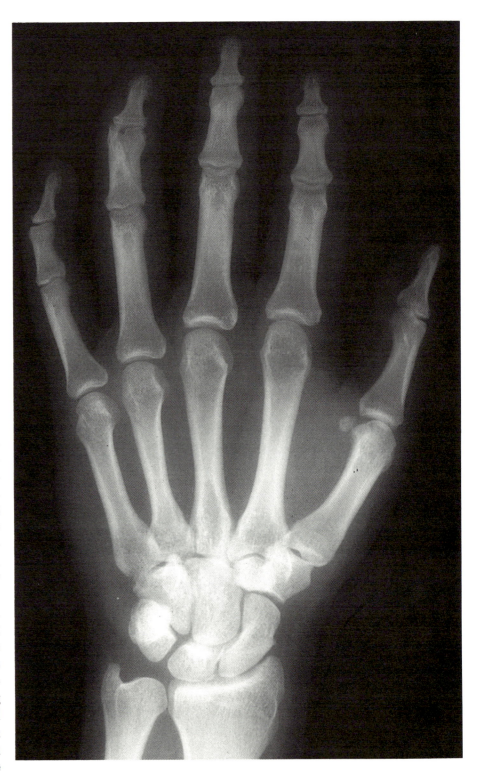

X-ray of a normal left hand.

A century later, scientists are now well informed about the advantages and disadvantages of the X-ray procedure, which does, of course, involve directing dangerous radiation at the human body. In small controlled doses this does little or no harm, and X-rays are widely used throughout the world to diagnose broken bones, teeth defects and countless other medical conditions. Hospitals and dentists take great precautions to ensure that patients are exposed to the absolute minimum dose of radiation necessary for the X-ray. X-rays are not taken unless they are considered essential: pregnant women are only X-rayed in great emergency, as the radiation may harm the unborn child. Without doubt the procedure is as safe as it can be, and X-rays contribute in a very positive way to medical diagnosis nowadays.

Procedures have also been developed for the positive use of radioactivity in helping to cure patients with life-threatening diseases such as cancer. The harmful cancer cells in their bodies can be killed by controlled doses of radiation. Such treatments, known as radium therapy, must be strictly measured and delivered so that the surrounding healthy cells in the body are not damaged.

Although many people died in the early days of X-rays through the indiscriminate use of a process that was not fully understood at the time, there can be no doubt that Roentgen's discovery has saved countless lives since. Radiation will continue to be a great asset in the diagnosis and cure of disease well into the next century.

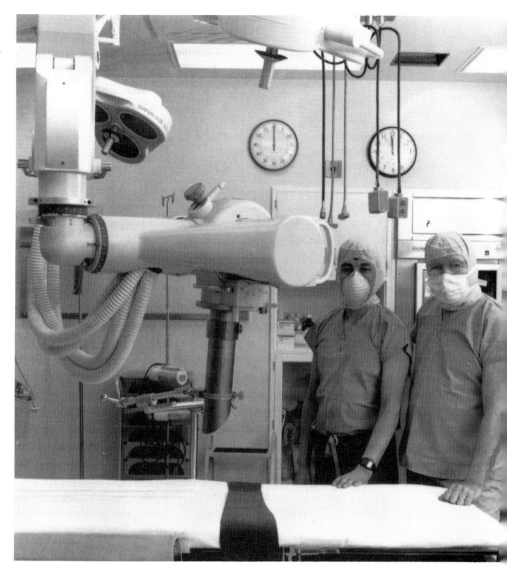

Stanford University Medical Center, California, USA, at the forefront of research and development. This unit uses a technique known as intraoperative radiotherapy. The treatment is given to cancer patients inside a specially-equipped and lead-shielded operating room at Stanford University Hospital. It is effective for patients with cancer deep inside their bodies. The tumour is surgically exposed and radiation can be directed at it, avoiding normal tissues and organs. A single treatment will provide six to eight times the normal daily dose of radiation. This specially equipped room began its service in 1990.

FOOD IRRADIATION

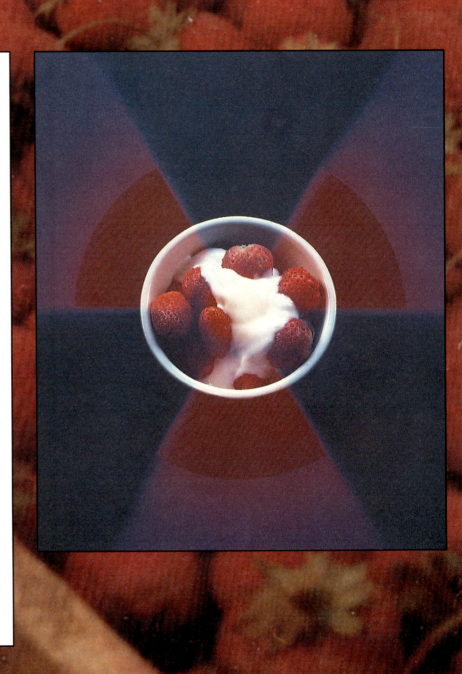

A new method of treating food known as irradiation has proved highly controversial. In the minds of many people irradiation conjures up alarming thoughts of disease and danger. The prospect of combining radiation with the food we eat cannot be dismissed lightly, though it is necessary to have a much better knowledge and understanding of the irradiation process before deciding our attitude to the government's plans.

All food is naturally radioactive to some extent. This level of background radiation in food is no more dangerous than that found in our own bodies or the homes we live in. So why irradiate, and what exactly does this mean?

The original reason behind irradiation was to make food last longer, thus preserving its shelf-life in shops and reducing the number of food poisoning cases. During the procedure food is passed through a concrete-lined chamber where it is treated with a carefully measured amount of ionizing radiation, either by using gamma rays or by electrons from an electron accelerator.

ÚJ ELJÁRÁS — ÚJ TERMÉK
GRILLCSIRKE
SUGÁRENERGIÁVAL KEZELVE

AZ ELJÁRÁS ELŐNYEI:

VÁSÁROLJON ÚJ TERMÉKÜNKBŐL !

Food irradiation

THE GOOD

Food keeps longer
Increases shelf life
Cuts down on waste
Reduces food poisoning

THE BAD

Cost increases
Damage to vitamins in food?
Not all bacteria are destroyed, so
might give false sense of security
Psychological ... conjures up a
negative idea

The amount of radiation required to irradiate food varies, depending on the density and the texture of the food. As a result of irradiation treatment, cells within the food are changed with interesting results. For example, irradiation delays the ripening of fruit, it kills off insect pests and reduces the number of micro-organisms that cause food to decay. Scientific evidence suggests that the dose of radiation is not sufficient to make the food dangerously radioactive, though many consumers will no doubt continue to hold their own more cautious views.

The strong argument in favour of irradiated food is that it is safer for the consumer, as it contains fewer microbes and signs of mould and decay. Shopkeepers and retailers are able to buy larger quantities of food that will keep for longer, thus reducing their transport and handling costs.

Irradiation is not new, and early investigations were carried out in the United Kingdom as long ago as 1930. Today many scientists are convinced that it is a safe and valuable process. Indeed, as a method of food processing, it has been more exhaustively tested than any other. In 1970 an international food irradiation project began in which 24 countries participated. This project continued for over a decade and concluded that the irradiation process was not dangerous. It is approved by the World Health Organization, the United Nations' Food and Agriculture Organization and the Food and Drug Administration of the USA.

In recent years, the number of food irradiation plants has increased rapidly throughout the world, and it is currently (1992) practised in over twenty countries. In 1988 the USA had three food irradiators and five under construction or planned; Russia irradiates 400,000 tonnes of grain a year; whilst Japan irradiates 20,000 tonnes of potatoes.

One major concern which has been expressed about irradiation is the question of whether the nutritional value of food is affected. The London Food Commission points out that damage to a wide range of vitamins occurs. Others argue that nutritionally there appears to be little change in irradiated food.

At the time of writing, food irradiation is not legal in Britain. When it becomes so, it is certain that the UK government will organize strict controls. These include plans to have a licensing

Irradiated foods are widely sold in a number of countries. In Hungary, for example, some 23 kinds of irradiated food products are advertised for sale.

system for treatment plants and a body of inspectors who regularly visit these plants to investigate safety levels. Comprehensive records will be kept of all irradiated food, and details will be displayed on the packaging. Finally, the public can be sure that not all food will be irradiated, and that labelling in the shops will leave the choice of whether to buy or not to buy for those in doubt.

How can irradiated foods be identified when on sale? It is up to each individual country to decide on a wording or a symbol. The symbol above, however, is widely accepted as an identification label for food that has been treated with ionizing radiation.

Strong reassurance

The UK Ministry of Agriculture and Food issues strong words of reassurance to anyone doubting the safety of irradiated food. It dismisses the misconception that scientists are themselves divided on the subject, assuring us that 'safety has been accepted by every independent scientific committee that has ever considered the subject.'

It is very good news that extensive research also gives no evidence of cancer risk or of vitamin losses.

With regard to increased radioactivity, it is pointed out that all food is naturally radioactive and that the very minute amount of extra radioactivity involved in the irradiation process is so small that it isn't even measurable by scientific instruments. Indeed the EC Scientific Committee for food calculated the increase to be 100,000-fold smaller than levels that occur naturally in food.

Furthermore, research shows that irradiation is very effective against the main causes of food poisoning.

Irradiation may be regarded as part of a comprehensive food safety strategy.

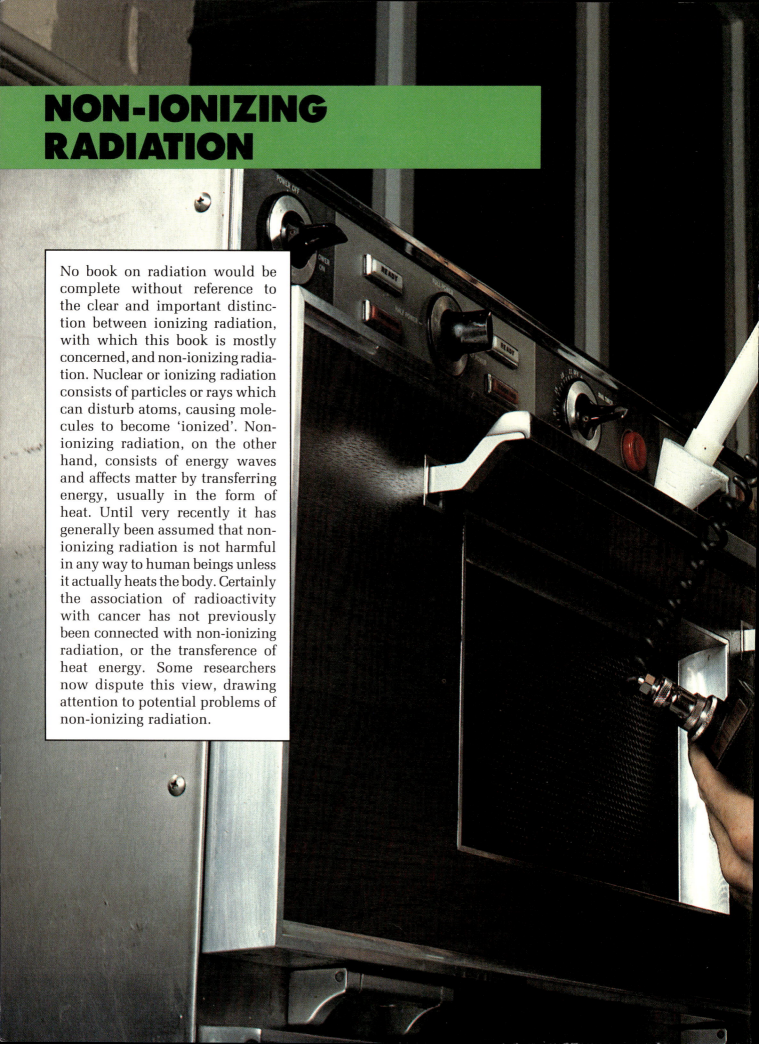

NON-IONIZING RADIATION

No book on radiation would be complete without reference to the clear and important distinction between ionizing radiation, with which this book is mostly concerned, and non-ionizing radiation. Nuclear or ionizing radiation consists of particles or rays which can disturb atoms, causing molecules to become 'ionized'. Non-ionizing radiation, on the other hand, consists of energy waves and affects matter by transferring energy, usually in the form of heat. Until very recently it has generally been assumed that non-ionizing radiation is not harmful in any way to human beings unless it actually heats the body. Certainly the association of radioactivity with cancer has not previously been connected with non-ionizing radiation, or the transference of heat energy. Some researchers now dispute this view, drawing attention to potential problems of non-ionizing radiation.

A common source of this form of radiation in our homes is the microwave oven. These machines are used to cook food, or to reheat previously cooked foods, very quickly. The electronic or microwave energy they generate is a type of high frequency, invisible radio wave. Unlike the traditional oven, there is no direct source of heat applied to the food. Such traditional cooking is a slow process whereby the heat energy of the oven is gradually absorbed by the food. Microwaving, on the contrary, is a far speedier operation.

As microwaves pass through to the food in the oven, they cause molecules of water within the food to vibrate at extremely high speeds. This friction causes intense heat which rapidly cooks the food. A great advantage of microwave ovens is that they can cook in a quarter of the time taken by traditional methods. It is also argued that foods cooked by microwaves actually taste better and are better for us because there is less time for dishes to lose their vitamins and flavour.

Microwave energy will not pass through all substances: some will absorb the waves or reflect them. It is important for cooks to bear this in mind when choosing

Electromagnetic spectrum

long wavelength · · · · · · · · · · · · · · · short wavelength

| RADIO WAVES | MICROWAVES | INFRA-RED | VISIBLE LIGHT | ULTRAVIOLET | X-RAYS | GAMMA RAYS |

low energy · high energy

Energy Waves of the Electromagnetic Spectrum. This is a method of classifying waves of energy that are radiated from various sources. Such waves include radio waves, microwaves, light, x-rays and gamma rays.

utensils for use in microwave ovens. Metal reflects microwaves and so metal utensils cannot be used in this form of cooking. Glass, china, plastic, wood, wicker and paper utensils, however, allow the waves to pass through and go directly to the food that will absorb them.

Microwave ovens are made to strict safety standards and microwaves cannot escape through the sealed door of the oven. Scientists believe that these ovens are completely harmless unless they are faulty or damaged. Nevertheless, others remain cautious about their safety, suggesting that microwaving affects the food adversely so that unwanted electromagnetic energy is passed into the body when the food is eaten.

Measuring radio waves around an induction furnace.

Other sources of non-ionizing radiation include television sets, visual display units (VDUs) and radar systems. There has been considerable publicity about VDUs. Many workers who spend long periods of time in front of computer monitors or screens have complained of a range of problems including tiredness, headaches, eye strain and anxiety. There is also concern about pregnant women using them. VDUs can be fitted with anti-glare screens which may ease eye strain problems, but will not counteract the radiation that is emitted. Much research is currently being undertaken in this area, and stricter safety measures relating to the controlled use of VDUs will no doubt be introduced in due course.

It must be remembered that radiation in our homes is part of our natural environment. It is important to come to some understanding of radiation and to develop an informed awareness of its potential dangers. Everyone can take steps to protect themselves, and pressure can be put on industry and governments to adopt strict safety and monitoring procedures.

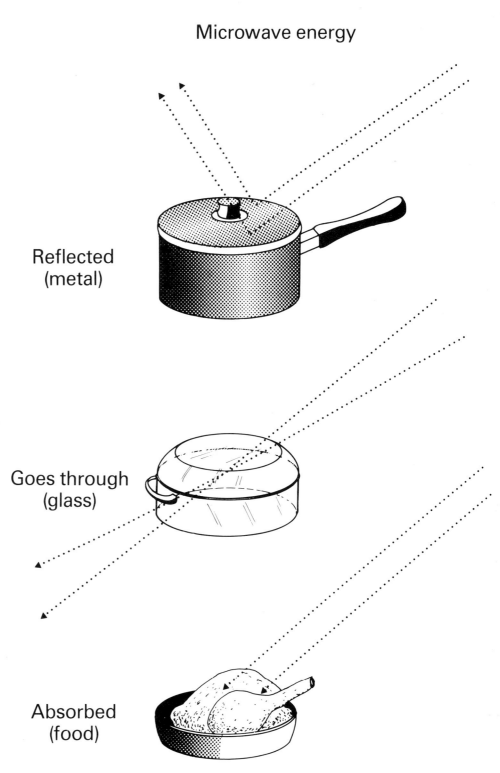

Microwave energy

Reflected (metal)

Goes through (glass)

Absorbed (food)

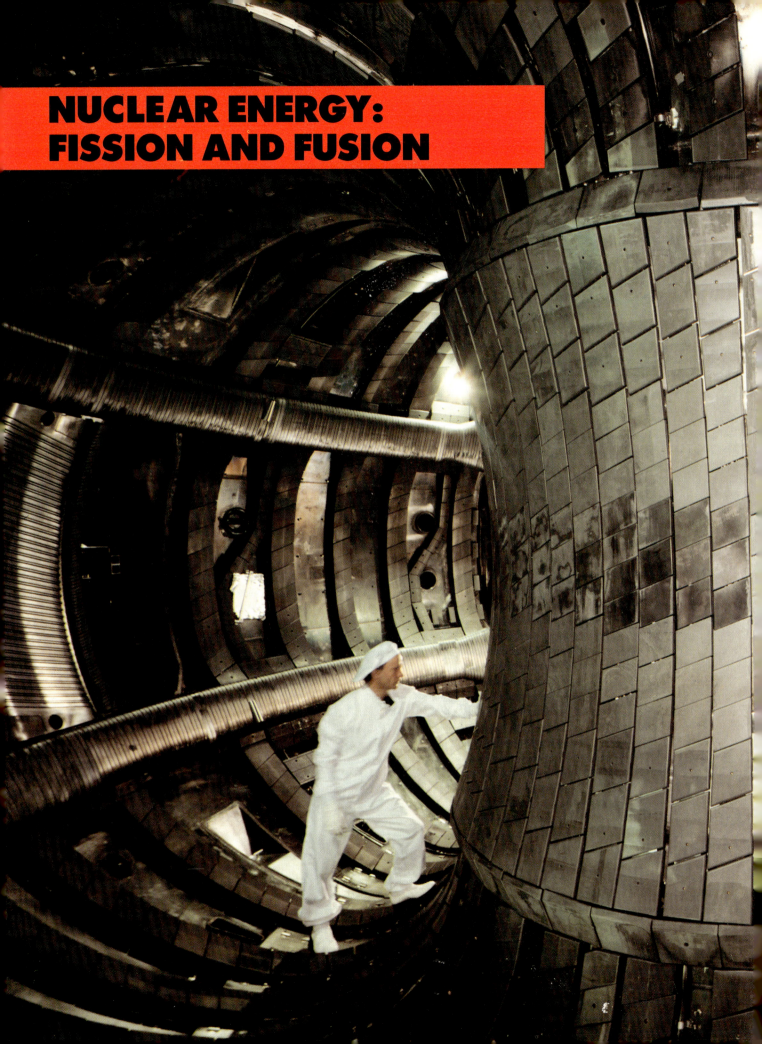

NUCLEAR ENERGY: FISSION AND FUSION

One method by which nuclear energy can be released is through a process known as fusion. Very light atoms such as hydrogen have to be heated to an enormous temperature of hundreds of millions of degrees Centigrade. At such temperatures, the nuclei are moving so fast that they join or 'fuse' together: as the nuclei fuse, the protons and neutrons making up the nuclei combine to form larger nuclei, generating huge amounts of energy. The sun's own energy is produced by this fusion process.

At the present time there are no power stations using nuclear fusion, since scientists have not yet achieved simultaneously all the necessary conditions for atoms to fuse, although they have obtained the required temperatures and other conditions in separate experiments. Teams of scientists throughout the world are working to try and produce controlled nuclear fusion in the laboratory, using specially constructed experiments to create the required temperatures and methods of containing very hot gases (called plasmas) with magnetic fields.

In March 1989, newspaper headlines excitedly announced that scientists in America reported success in controlling nuclear fusion. A team working at Utah University claimed to have produced controlled fusion in a test tube at room temperature. Many scientists considered this to be a dubious claim, and numerous teams throughout the world have tried in vain to repeat Utah's experimental results. At the time of writing, there is no conclusive evidence that energy can be produced in this way on a large scale.

The world's leading fusion research machine is known as the Joint European Torus (JET fusion machine), located at Culham near Oxford in the UK. The JET project is funded by and has staff from fourteen European countries. Experiments in JET have achieved, for a few seconds, temperatures up to 300 million degrees Centigrade – about twenty times hotter than in the centre of the sun. The next stage after JET is to build an experimental reactor to harness the fusion energy and thereafter a fusion power station. This will not be until well into the next century. Similar machines are located in the United States of America, Russia and Japan. Scientists are anxious to solve the fusion problem because success in developing this process would end all worries of fuel for energy supplies in the future. Furthermore, it would end the serious concerns the world has about the dangers of radioactivity associated with nuclear fission.

Nuclear fission is the main method of producing nuclear energy. It involves 'splitting' atoms. At the centre of an atom are two groups of particles called neutrons and protons. These are within the nucleus. Not all atoms are the same: some have different numbers of protons from others. Outside the nucleus are moving particles known as electrons which spin around the nucleus, rather like planets around the sun.

The Joint European Torus. The two men give an idea of the scale of this massive machine.

Fission

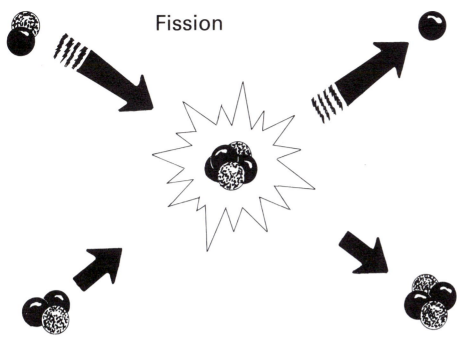

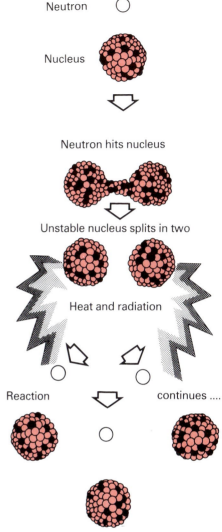

Neutron

Nucleus

Neutron hits nucleus

Unstable nucleus splits in two

Heat and radiation

Reaction   continues ....

The fission reaction of deuterium and tritium gives a nucleus of helium and an energetic neutron, that is, a neutron that generates electricity.

In 1939 the discovery was made that some uranium atoms can be 'split'. 'Free' neutrons moving around the uranium fuel start the fission process. When an extra neutron is added to the nucleus of an atom of fuel it causes this nucleus to break into two. As it splits, more neutrons are given off or set free. These freed neutrons can in turn cause more fission of other nuclei and produce even more neutrons ... and so on. This continuing process is called a chain reaction.

Freed neutrons and split nuclei move at very great speed which generates heat. This heat is the nuclear energy produced by the fission process. Very small amounts of fissionable fuel are capable of generating colossal amounts of heat.

The type of uranium atoms which can be split are called uranium 235 (U-235) atoms. If an extra neutron strikes the nucleus of a U-235 atom, this splits and the chain reaction gets under way. When fission is used to produce nuclear power, the process is controlled so that the chain reaction proceeds at a steady pace and heat is constantly produced.

THE NUCLEAR INDUSTRY

Nuclear industry and power is a complex and very controversial subject. The industry not only generates electricity but also creates a range of products for use in hospitals, agriculture and other industries. Nuclear power is taking on an ever-expanding role in our world, and our energy needs are increasingly being met by nuclear power. In Britain there are 18 working power stations which between them produce just over 20 per cent of the total electricity used. In 1987 the USA had 106 nuclear power reactors in existence and another 13 under construction. Nuclear plants supply 18 per cent of the USA's electricity.

The proportion of power derived from nuclear energy is growing. Many believe that this trend must continue, replacing the more traditional ways of producing electricity by burning coal or oil, fuels which cause more pollution and may eventually run out. Others are of the opinion that the proportion of the world's electricity supplied by nuclear plants will actually decline. The expense of building and maintaining nuclear power stations has risen steadily. In the USA, recently-built plants generate power at a cost of more than 13 cents per kilowatt hour, which is twice the rate of other forms of electricity.

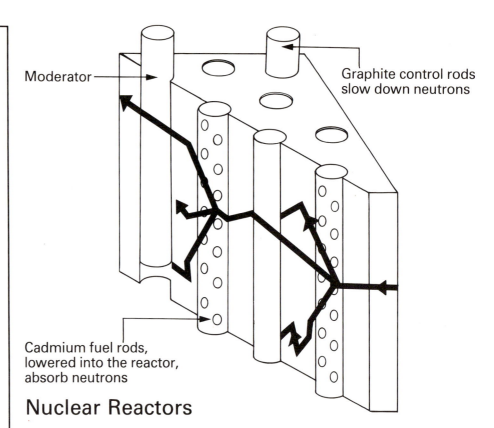

Moderator

Graphite control rods slow down neutrons

Cadmium fuel rods, lowered into the reactor, absorb neutrons

Nuclear power stations: for or against?

FOR

Clean: no toxic fumes to pollute the atmosphere
No ash
No lorries or trains bringing coal
Plenty of fuel available

AGAINST

Danger of radiation – not confined to the local area or even to the country owning the power station
Radioactive waste – disposal problems
Some waste could be used for making bombs
Dangers of accident or terrorism when transporting fuel or waste

Nuclear Reactors

Many new orders for nuclear plants in the USA have been cancelled because of increasing costs. No new orders for commercial reactors have been placed in the US since 1974.

Nuclear power stations use a fuel called uranium. This is a mineral found in certain types of rock such as granite. A great difference between nuclear and coal/oil powered stations is that in the former, the fuel is not burned. Nuclear power is produced by 'splitting' atoms or 'fission', a process explained in detail on the preceding pages.

Uranium is found in rocks in certain parts of the world including Australia, the United States of America, Canada and North Africa. It is mined and then separated from the rock to produce a substance called uranium oxide which is exported to countries who require it for nuclear power stations.

Nuclear power stations contain nuclear reactors. In a reactor are fuel rods containing the uranium or other fuel capable of fission, a moderator and control rods. Heat is produced by the splitting (fission) of the nuclei of atoms of the uranium fuel. The process also produces neutrons – rapidly moving particles which can collide with other atoms and cause them to split producing further neutrons which cause further fission. This is the chain reaction.

The function of the moderator in the reactor is to slow down the neutrons in order to help fission occur in the fuel rods. The quantity of neutrons available to cause further fission is regulated by the control rods which are made of boron steel, a material which absorbs neutrons and thus slows down or stops the chain reaction. During the fission process, the reactor becomes hot and the heat is extracted by a coolant liquid or gas which is pumped through the reactor core. Next the heated coolant passes through a heat exchanger where it causes water to reach boiling point. Steam produced as a result then powers a steam turbine in much the same way as a boiler does in a traditional fuel-burning power station.

In the United Kingdom there are several kinds of nuclear reactors. Magnox, in which the nuclear fuel is in magnesium alloy cans and the coolant is carbon dioxide gas, is the oldest type; AGR (advanced gas-cooled reactors) are similar but operate at a higher temperature; PWR (pressurized water reactors) use water as a coolant, but at such a high pressure that it does not boil. Fifty

Pressurized Water Reactor

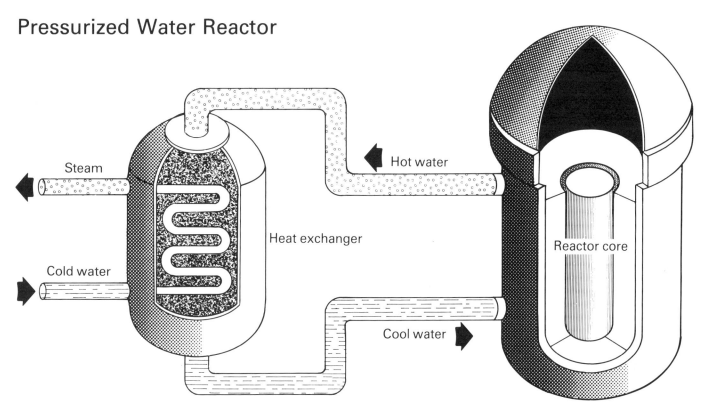

per cent of the world's nuclear reactors are PWRs.

Inevitably there are arguments for and against the use of nuclear reactors to produce power. They are clean and use a fuel which is readily available. Furthermore, they are efficient in their use of fuel and huge amounts of energy are released from it: for example, one kilogram of uranium can produce a staggering two million times as much energy as one kilogram of coal. The main worries are about their safety and these are discussed later. Another key argument concerns the cost of building and maintaining nuclear power stations, as mentioned in connection with cancelled plans in the USA.

Control room at Calder Hall nuclear power station in West Cumbria.

HOW SAFE IS NUCLEAR POWER?

The process of nuclear fission for generating power is very efficient. It is also considered to be highly dangerous. It produces very large amounts of radiation and also radioactive waste. As we already know, there are different levels of radiation. We are surrounded by a background of low-level natural radiation throughout our lives, which generally causes little harm.

The production of nuclear power, on the other hand, causes a far more intense type of radiation known as high-level radiation. As the atoms split, fast-moving particles and rays are emitted. Within a power station, therefore, the fuels themselves are highly radioactive and therefore hazardous. Plutonium, a commonly used fuel, could be lethal if its dust is breathed in. Even a tiny amount of inhaled dust may cause lung cancer. This raises serious issues relating to standards of health and safety for those who work within nuclear power establishments or who live nearby.

As far as employees are concerned, the utmost care is taken not to put them at risk. Protective suits are always worn in areas of the power station where radiation may be present. After being on duty, all workers are 'checked' – their radiation levels are measured by a body monitor. The air within the building is constantly monitored by accurate instruments. Radioactive material is never handled directly, but kept in concrete or steel-lined containers which are completely leak-proof. If material does have to be handled in the station laboratory or elsewhere, this is never done with bare hands but always by workers equipped with protective gloves.

TRANSNUCLEAIRE

ACLU 368534

MA. 80

The entire power station must, of course, be leak-proof for the safety of those living nearby. The atmosphere outside is constantly checked for raised levels of radiation. Unfortunately we too often hear of power station 'leaks' and of serious hazards associated with them.

One very problematic area associated with the nuclear power industry is the disposal of the waste generated by the process. This waste varies in terms of its danger level but includes spent fuel, which is highly radioactive and extremely dangerous. Waste can be in solid, liquid or gaseous form, and is classified into three groups, depending on the level of its radioactivity. These groups are: a) low-level waste; b) intermediate-level waste; and c) high-level waste.

The way in which waste is dealt with depends on its classification. Low-level waste from power stations, hospitals and other industries – indeed from any establishment where ionizing radiation occurs – is relatively harmless. Examples of this waste include items such as protective clothing, towels, medical supplies, paper from laboratories and general radiation-contaminated industrial rubbish. Much of this is considered so harmless that it is disposed of fairly easily, for example, by placing it in deep trenches dug in the ground which are then filled with a covering layer of at least one metre of soil.

In the UK low-level solid wastes are dealt with in this way near the Sellafield nuclear power station in Cumbria. The site is fenced off and radioactivity in the area is

A modern industrial robot at work inside the Oak Ridge nuclear installation in the US.

This whole body monitor is used to measure radioactivity in volunteers and also in people contaminated at work.

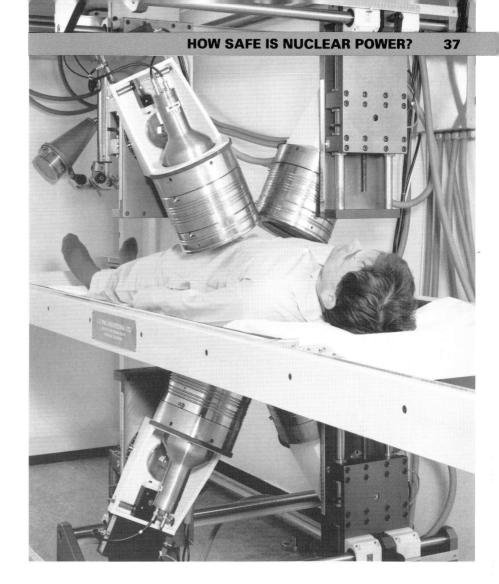

regularly monitored to ensure that a dangerous level is not reached. In the past, commercial sites have been used in the USA to bury low-level wastes. Three such sites, however – at Barnwell, South Carolina; Beatty, Nevada and Hanford, Washington – are closing, either because they are full, or for safety reasons. A 1980 Act of Congress mandated that all states should be members of a compact to dispose of radioactive waste or build their own disposal facilities by 1990.

Other low-level wastes are dumped in the oceans. A common practice is to contain the wastes in steel drums. Within the drums, the actual waste materials are enclosed in concrete. The drums are then taken to the coast and cast into the sea.

Lower-level liquid wastes may be diluted and discharged into rivers, lakes and the sea. This is of course highly controversial as many people believe that seaweed, water, fish and shellfish are inevitably subjected to radiation which could then in turn find its way into human food chains. Water samples are regularly analysed to check for rising levels of radioactivity.

An equally serious matter is the practice of discharging low-level gaseous waste into the atmosphere. Scientists believe that any individual dose of radiation generated in this way is so small as to be harmless, though many people, especially those living close to nuclear power stations, are naturally very concerned at this situation.

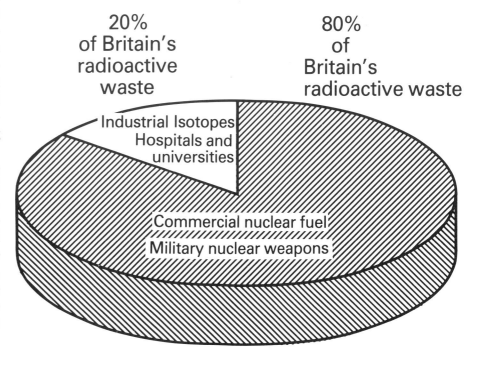

20%
of Britain's
radioactive
waste

80%
of
Britain's
radioactive waste

Industrial Isotopes
Hospitals and
universities

Commercial nuclear fuel
Military nuclear weapons

INTERMEDIATE AND HIGH-LEVEL WASTES

Intermediate-level wastes from nuclear industry have to be dealt with in a more complex and secure way than low-level wastes. In the United Kingdom and the USA the main method for their disposal is to bury them in deep underground trenches. Wastes that will remain radioactive for a very long time are buried even deeper. Intermediate-level wastes may include metal cans that have been used to contain nuclear fuel in a power station reactor, clothing and equipment that has been contaminated to a significant degree, and sludges derived from the treatment processes in nuclear power stations.

Overall responsibility for dealing with radioactive materials lies ultimately with the government but, in practice, disposal is generally left to the organizations and plants which produce the waste. In Britain, an organization called NIREX (the Nuclear Industry Radioactive Waste Executive) was established in 1982, and reports regularly to the government. NIREX is responsible for overseeing the disposal of all Britain's intermediate- and low-level solid wastes. All nuclear waste disposal must adhere to extremely strict regulations, and is subject to constant scrutiny and monitoring; it must only be carried out with government permission.

Positioning of a waste flask on to a tipping machine in the highly active waste silo building at Sellafield, Cumbria.

High-level radioactive waste comes from only one source; the reprocessing of nuclear fuel which has already been used. The fuel consists of atoms of uranium or plutonium. Used fuel is certainly not just waste – it contains substantial quantities of uranium (96 per cent) and also of plutonium, both of which can be separated from the waste and re-used. The valuable plutonium and uranium are removed and separated by dissolving the used fuel in acid. What is left after this process is extremely radioactive and emits high levels of heat: it must therefore be cooled and handled with extreme caution.

Most high-level waste is in liquid form. It is stored in stainless steel tanks which are surrounded by concrete. At Sellafield in Cumbria there are 19 of these tanks, which contain all the high-level waste from the entire nuclear industry of the United Kingdom. The quantity of high-level waste of this kind is not very great: the entire UK nuclear programme has generated only around 1800 cubic metres of liquid in the last 30 years. It is nevertheless extremely dangerous and must, at all costs, be prevented from escaping into the environment. In the USA, as in Britain, the most serious disposal problem concerns the high-level waste and fuel rods from nuclear reactors. 15,000 tons of this fuel are stored in cooling ponds in the 106 reactor sites. The US Nuclear Waste Policy Act states that these substances must be stored in a geologically stable location 1,000 to 4,000 feet underground. A planned site at Yucca Mountain, Nevada, would hold

Nuclear waste disposal is a controversial subject about which every reader will have personal opinions. Questions relevant to conservation and which are constantly debated include:

- Is it right to dispense radio-activity freely into our environment, no matter how 'low level' it is claimed to be?

- How can we be assured of the long-term and cumulative effects of such disposal?

- Just how safe are under-ground and ocean disposal methods?

- Can we justify using nuclear energy at all in our world when its potential danger to life is so serious?

70,000 tons of waste for 10,000 years. It is expected to cost $100 billion which does not include temporary storage while the site is being constructed.

It is believed that high-level nuclear waste can be managed more easily if it is in solid blocks rather than liquid form. It can be turned into glass blocks which can then be stored for many years, until the heat and radioactivity have lessened, and then finally disposed of.

The process of solidification and incorporation into glass is 1known as vitrification. The vitrified waste is enclosed in containers made of stainless steel, which are then in turn surrounded by reinforced concrete. Some countries have already started using this process; in France, for example, it has been practised since 1978 and in Britain since 1990.

Tank for storing high-level liquid waste

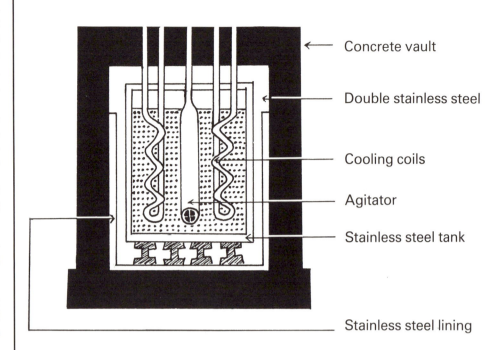

- Concrete vault
- Double stainless steel
- Cooling coils
- Agitator
- Stainless steel tank
- Stainless steel lining

Vitrification

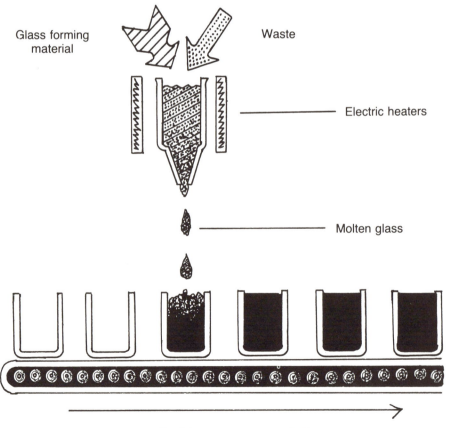

Glass forming material

Waste

- Electric heaters
- Molten glass

Containers on conveyor belt

There are two possible ways of disposing of the high-level solidified waste: it can either remain in suitable places deep under the ground, or be disposed of on the ocean beds. These ideas seem to be the most practical and sensible, although other considerations have included sending the waste into space by rocket, or burying it in the ice caps of the polar regions. Whether the waste ends up on land or at sea, the ultimate consideration must be whether life on earth is sufficiently protected from any danger. The barriers between the waste and the rest of the world must be strong, and durable enough to prevent significant quantities of radiation from reaching our atmosphere and environment. The waste itself must also be in a lasting and stable form, so that no leakage occurs. Extensive testing and research have shown vitrification to be a safe process, as the glass blocks are both stable and durable.

DISPOSAL OF HIGH-LEVEL WASTE

UNDERGROUND?

SAFETY BARRIER = CONTAINERS + ROCK

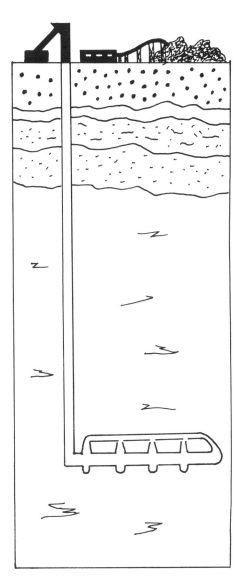

Many countries have research programmes investigating the effectiveness of burying waste deep in the ground.

Barriers to Waste would be those encasing it (i.e. constructed and engineered by people) plus the rock of the earth itself which would form the waste's ultimate surroundings.

ON THE SEA BED?

SAFETY BARRIER

= SEDIMENT + DILUTION

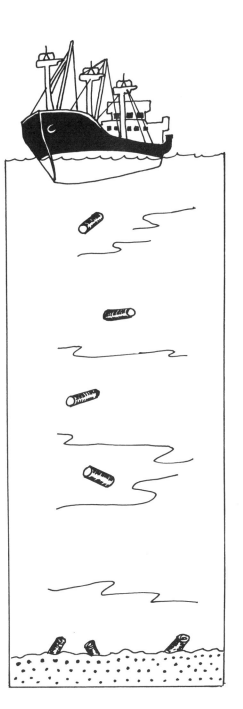

On the ocean floor in some parts of the world are deep accumulations of sediments that have been present for millions of years. These would form an effective barrier between the waste containers and the ocean environment. If any radioactivity did leak and reach the surface of the sea bed, then it would be diluted to such an extent as to render it harmless.

WHEN THINGS GO WRONG

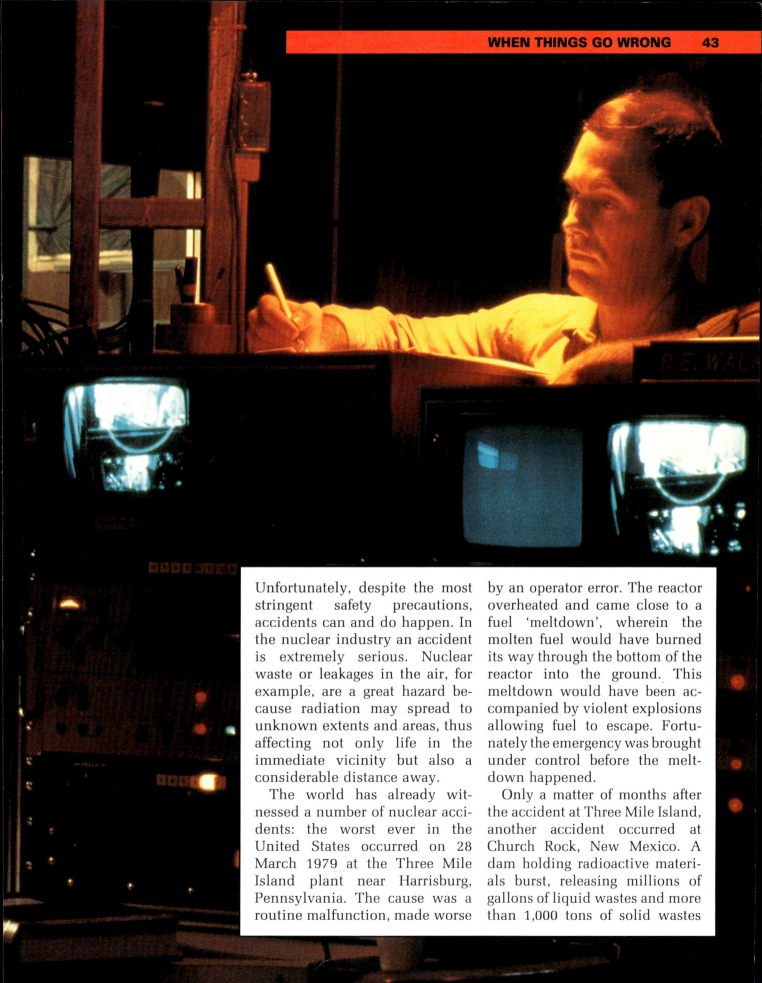

Unfortunately, despite the most stringent safety precautions, accidents can and do happen. In the nuclear industry an accident is extremely serious. Nuclear waste or leakages in the air, for example, are a great hazard because radiation may spread to unknown extents and areas, thus affecting not only life in the immediate vicinity but also a considerable distance away.

The world has already witnessed a number of nuclear accidents: the worst ever in the United States occurred on 28 March 1979 at the Three Mile Island plant near Harrisburg, Pennsylvania. The cause was a routine malfunction, made worse by an operator error. The reactor overheated and came close to a fuel 'meltdown', wherein the molten fuel would have burned its way through the bottom of the reactor into the ground. This meltdown would have been accompanied by violent explosions allowing fuel to escape. Fortunately the emergency was brought under control before the meltdown happened.

Only a matter of months after the accident at Three Mile Island, another accident occurred at Church Rock, New Mexico. A dam holding radioactive materials burst, releasing millions of gallons of liquid wastes and more than 1,000 tons of solid wastes into the Little Puerco River. This was potentially the worst incident of radiation contamination in the history of the USA, but fortunately it occurred in an isolated part of the continent.

Britain has also been a location for nuclear accidents: in 1986, for example, it was reported that on one occasion 15 tonnes of mildly radioactive carbon dioxide gas escaped into the atmosphere from the Trawsfynydd nuclear power station in North Wales. Fortunately this was a minor incident, with no threat to life; only a small area of ground outside the boundary of the power station was found to be slightly contaminated with radioactivity.

Another location in the United Kingdom where safety is regularly questioned is the Sellafield nuclear processing plant (formerly known as Windscale) on the west coast of Cumbria in the north of England. As well as high-level waste, it produces around 70 per cent of all low-level radioactive waste in Britain, and 40 per cent of the intermediate-level wastes. All reprocessing of used nuclear fuel is carried out at Sellafield, where plutonium and uranium are separated out for re-use.

How safe is Sellafield? Might radioactivity affect the lives of people living close by? Could it harm the environment? There are no easy or straightforward answers to these important questions. Reports about Sellafield give cause for grave concern. This page provides some facts, information and ideas which, it is hoped, will be helpful to anyone considering this serious issue. No matter how 'absolutely safe' contained radioactive material might be, two facts remain certain: firstly, accidents can and do occur, no matter how stringent the safety precautions, and secondly, we can never be sure that we understand a situation totally – perhaps there are hidden dangers we do not know about.

The entrance to the damaged nuclear plant at Three Mile Island in the US.

Monitoring a transport flask containing irradiated nuclear fuel.

An example of this second point is that, since 1983 when media attention was drawn to the problem, Sellafield has, rightly or wrongly, been blamed by many for the fact that local children seem to show a higher than average chance of becoming ill with the blood cancer leukaemia. Various statistics, including those gathered by local campaigns, suggest there is a link between developing leukaemia and living close to a nuclear plant. An official full-scale enquiry found no definite connection between blood cancer and radioactivity from Sellafield, and this seems to suggest that there is no cause for alarm. However, no other obvious reason for the unusually high number of cases of leukaemia in the children of the area was discovered, and so it would appear that the Sellafield connection has so far been neither proved nor disproved.

In 1983 an accidental dispersal of radioactivity into the adjoining sea caused serious problems, for instance, killing marine life, and provoked severe criticism when contamination was washed back into the land by tides. Beaches in the Sellafield area were closed for three months until they could be declared 'safe' from radioactivity.

This event raises serious questions. It is possible that a lot of dangerous waste washed into the sea is never drawn to our attention because it is not discovered, or because it cannot be proved to have come from Britain. Politicians, individuals and groups oppposed to nuclear power have

claimed that traces of radioactivity derived from Sellafield can be found in fish in waters as far away as Sweden. A report published by one group of MPs in 1986 claims that 'the Irish Sea is the most radioactive sea in the world' and that Sellafield is 'the largest recorded source of radioactive sea in the world'. Such

Sellafield in West Cumbria.

claims are obviously extremely serious, and are taken as such by those responsible for the nuclear industry. Controls on the discharge of waste are constantly monitored and reviewed, and research continues into safe methods of disposal.

CASE STUDY: NUCLEAR WASTE IN CUMBRIA

The county of Cumbria in England is an area at the centre of great controversy and public concern over nuclear matters, and not just because of the danger of leaking radiation and accidents at Sellafield. Of equal significance is Cumbria's role in the planned disposal of nuclear waste.

Much low-level waste is disposed of by burial in trenches at Drigg, near Seascale, and into the Irish Sea via a pipeline at Sellafield. In 1985 the Friends of the Earth organization declared that the Drigg repository leaked and expressed great concern, since a stream runs through the site into

the local marshland where increased levels of radioactivity have been measured. Sheep grazing on this marsh were found to have significantly higher levels of radioactivity in their bodies than sheep grazing on pastures elsewhere in the north of England.

To add to this controversy, Cumbria again made newspaper headlines in March 1989 when it was announced that Sellafield had been chosen as the most likely of two possible sites for a new underground nuclear waste dump. The other possible site named was Dounreay in the north of Scotland. NIREX announced that it wanted to carry out detailed investigations at Sellafield.

The Secretary of State for the Environment granted British Nuclear Fuels permission to sink a trial borehole on the site. This overruled the decision of Cumbria County Council, which in 1988 had refused planning permission for a development of this kind. The planned repository would be some 200–500 metres beneath Sellafield.

One of the key arguments put forward for using this site is that three quarters of the low-level and two thirds of the intermediate-level nuclear waste generated in Britain is produced at Sellafield. Disposal of this waste on site would remove not only the expense of transport but also the risk of accidents when moving the hazardous waste elsewhere. At the time of writing, investigations at the site are on-going.

An organization known as Cumbrians Opposed to a Radio Active Environment reacted very strongly to the news, claiming that:

the nuclear industry is planning to make Cumbria a scapegoat for this country's nuclear waste problem ...

and adding that:

The environmental movement believes that all types of nuclear waste should be stored in above-ground dry stores at the site of the creation of the nuclear waste. That would cut out transportation problems. It would be a safe way of dealing with nuclear waste and if future generations find a proper way of dealing with nuclear waste then at least the waste can be retrieved and dealt with, unlike dumping.

This is an interesting viewpoint to consider and raises another vital issue which no book on this subject can ignore – our obligations to future generations. Planning should always take account of what may happen in decades to come. It is estimated that by the year 2030 some 1.5 million cubic metres of low-level waste will have been produced. The *Independent* newspaper made an interesting comparison: the cubic capacity of Big Ben's tower is about 0.014 million cubic metres, so NIREX is faced with the task of burying a volume of waste equivalent to about 110 Big Bens – and that is only low-level waste.

If planning permission is granted for the Cumbria site to go ahead, building work would commence by 1996 and be completed by the year 2005. British Nuclear Fuels is set to mount a major publicity campaign to persuade people that the underground store will be safe. Whatever the truth of the matter, it is not just a problem for our own generation. Whatever happens in the future with regard to the United Kingdom's nuclear policy, today's waste now exists and will continue to do so. There is no doubt that generations to come will have to be extremely careful about the dangers of today's waste. It will remain radioactive for over 100,000 years.

Oxide fuels (AGRS and Water Reactors) will be reprocessed in the Thermal Oxide Reprocessing Plant (THORP) from 1992. The uranium and plutonium can be reused in new fuel.

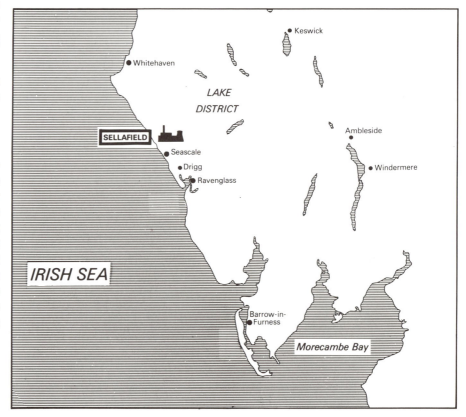

Sellafield in West Cumbria.

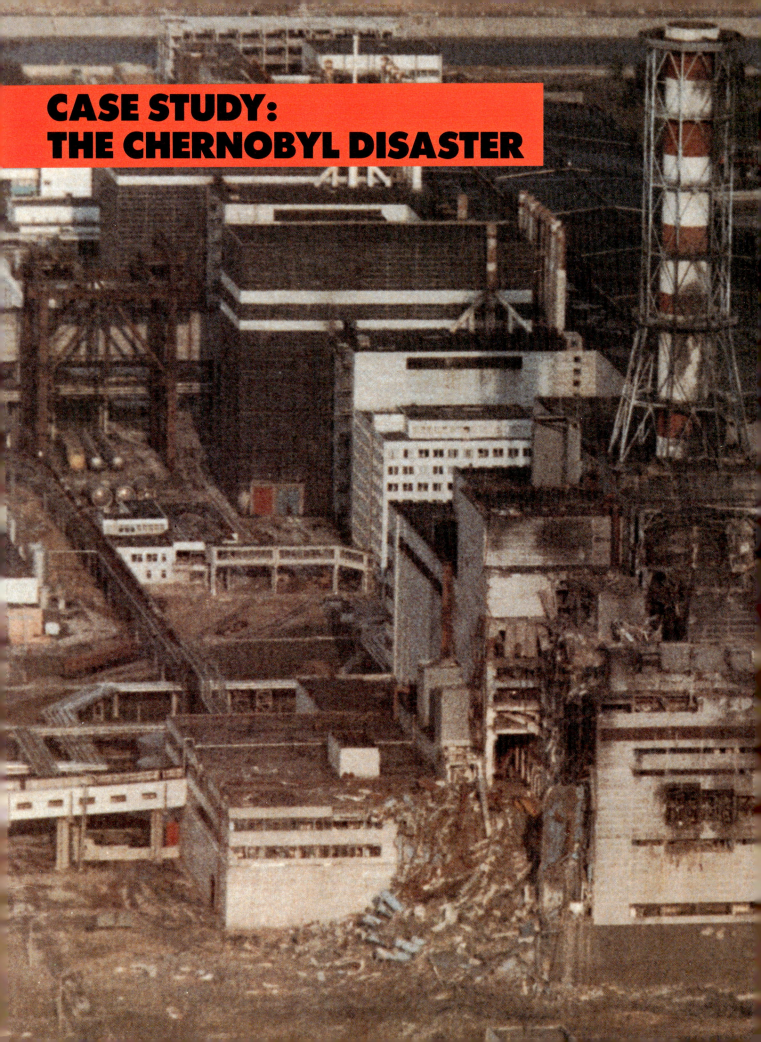

CASE STUDY:
THE CHERNOBYL DISASTER

The most serious incident caused by the accidental spread of radioactivity ever recorded was in April 1986, when there was a major radiation leak at the Chernobyl nuclear power plant, 128 kilometres (80 miles) north of the city of Kiev in the Soviet Union. This terrible event involved a tremendous fire in the nuclear reactor, known as a 'reactor meltdown'. The fire raged out of control and the effects brought suffering to people not only in the immediate vicinity but also in lands substantial distances away.

Three people were killed in the initial explosion at the reactor and, as a result of the accident, 237 victims were admitted to hospital, where 28 died within a short time. It may be impossible ever to calculate the final death toll: many survivors were allowed home after receiving treatment for radiation sickness, and their condition will continue to be monitored for many years. A tragic consequence of exposure to radiation can be the development of cancers, which might occur at any stage in an individual's future life. Indeed, although the long-term effects of the disaster on humanity may be predicted by scientists, they are far from certain. Many people were burnt by radioactive dust particles and suffered lung damage because of the radioactive smoke they inhaled; their future is bleak. Between 10,000 and 15,000 people were evacuated from the area in the immediate vicinity of the Chernobyl plant. All were considered to be at serious risk from the spreading airborne radiation. The land in the immediate vicinity will be unusable for many, many years.

Unfortunately, the risk was not confined to the immediate area. Two days after the explosion, a radioactive cloud had drifted more than 3,200 kilometres (2,000 miles) to Scandinavia where very high levels of radioactivity were reported in the atmosphere. Scientists throughout the world were monitoring the spread of the cloud, the speed and direction of the wind that carried it and the levels of radioactivity present in their own environments. Reports from Sweden announced air radiation levels ten times higher than normal, yet officials declared this to be 'harmless' and not high enough to pose danger to human life or food supplies. Nevertheless, the Swedish population was advised to take precautions, such as not eating meat from animals grazing in the area affected by the cloud, and to follow closely the constant media coverage of the cloud's progress as it passed over the country.

It was then reported that the radiation was moving north from Sweden towards the USA, but generally away from the UK. British scientists monitored the cloud, constantly advising on any increases in and potential danger from radiation in the UK. Although this danger was not significant at any point, many parts of the country recorded higher than normal radiation levels. Inevitably life was affected to some extent: many animals (mainly sheep) ate grass growing in affected areas, and this resulted in the production of meat that was considered unsafe to eat, as radio- activity can be passed between forms of life, along food chains. Because they could not sell their

Measuring rainwater for signs of radioactivity.

Chernobyl: direction of the radiation drift.

meat, many British farmers lost a lot of money as a result of the Chernobyl disaster.

In other lands, the effects were even more serious. In Lapland, for example, thousands of reindeer were contaminated by radiation. Many died. One hundred million people in Europe came under mandatory or voluntary food restrictions as the cloud spread and contaminated fruits, vegetables and grass for grazing livestock. The incident illustrates how any form of airborne waste/pollution may pose a tremendous threat to life throughout the world.

It would seem that this dreadful disaster was caused by a series of human errors, each successive one making the whole thing worse. The irony of this situation is that it was supposed to be a safety experiment which actually led to the catastrophe. The financial cost of the Chernobyl disaster to the USSR has been estimated as being eight billion roubles (about four billion pounds).

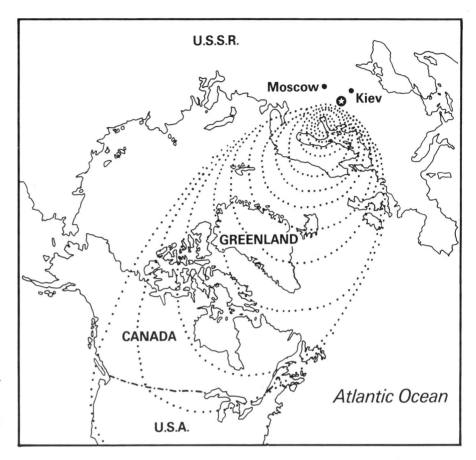

Monitoring sheep for radioactivity after the Chernobyl accident.

NUCLEAR WEAPONS

One of the most controversial issues in the world today is the use of nuclear energy for bombs, weapons and other military purposes. Nuclear weapons utilize the same fission process as power generation. In the nuclear reactor of a nuclear power station, the chain reaction is controlled so that it proceeds slowly; in a nuclear weapon, however, the chain reaction proceeds very quickly and the sudden release of intense heat produces a massive explosion.

The era of the 'atom bomb' dawned in a tragic and unforgettable way. The very first atomic bomb was tested in July 1945, exploding at Alamagordo in New Mexico. On 6 August 1945 a second bomb was dropped to explode on the city of Hiroshima in Japan. This was a uranium bomb, given the nickname of 'Little Boy'. A plutonium bomb, nicknamed 'Fat Man', was then exploded at Nagasaki. These are historic events that the world will never forget. Hundreds of thousands of people died as a result of the intense heat and radiation poisoning. Death was not only instantaneous: for weeks, months and years afterwards many more victims faced a lingering death caused by the radioactive fall-out.

Guide to nuclear weapons

BALLISTIC MISSILES:
These weapons are powered for the first part of their flight, then left to 'freewheel' to their target. They can travel extremely long distances and are often called Inter-Continental Ballistic Missiles. Usually, ballistic missiles leave the earth's atmosphere at some stage in their flight.

CRUISE MISSILES:
These are powered throughout their flight. Remote control or internal guidance systems will direct them to their target. They fly at much lower altitudes than ballistic missiles and are far less expensive. They can cause a nuclear explosion of up to 150 kilotons. This is the equivalent of 12 × Hiroshima.

Nuclear weapons may be termed STRATEGIC, INTER-MEDIATE or TACTICAL. These terms refer to their range and use.

STRATEGIC Weapons = large weapons, range over 5,500 km, launched from bombers or submarines.

INTERMEDIATE Weapons = weapons that range over 500 to 5,500 km. It was decided in the Intermediate Nuclear Forces Treaty signed by USA and USSR in 1987 to destroy all ground launched intermediate weapons by 1991.

TACTICAL Weapons = weapons that include short-range missiles, artillery shells, free-fall bombs and mines.

Since 1945 there has been a frightening escalation in the quantity and destructive power of nuclear weapons in the world. Present statistics are staggering. In 1945 the world had three atom bombs. Today there are some sixty thousand nuclear weapons spread around the globe. It is estimated that they contain an unimaginable sixteen million kilotons of explosive energy – **enough to kill everyone on earth twelve times over**. A sobering thought.

Even one single nuclear weapon has formidable power. Scientists estimate that one large bomb dropped on a city the size of Birmingham, UK or San Jose, California, would immediately kill two million people. Hundreds of thousands more would die later from burns, radiation poisoning and other injuries. It is actually very difficult to imagine or estimate the damage that would be caused. As well as the actual explosion, there would of course be a major fire raging through the city. The long-term effects are almost impossible to predict, but without doubt there would be no human or animal life surviving over a vast area surrounding the site of the explosion.

The Chernobyl accident helps to give some idea of the scale of such a disaster. In terms of its force, the energy equivalent of the Chernobyl disaster was 0.1 kiloton: in a radius of 29 kilometres (18 miles) 135,000 people were exposed to radioactivity as the cloud spread and scattered. Vast areas of land will remain uninhabitable for up to a hundred years. Yet – the power involved in the Chernobyl disaster is minute when compared to that of exist-

ing modern nuclear weapons. A single 'MX missile' contains 6,000 kilotons of explosive power. Even a limited nuclear war would bring incalculable damage to the world.

Quite apart from the death and devastation, there would be so much smoke, dust and debris in the atmosphere that the sun would be obliterated. The world would be in darkness, all day and all night, for many months ... temperatures would fall dramatically ... essential biological processes such as photosynthesis

that had survived the would inevitably die. This ful scenario is known 'nuclear winter'. It could gered with less than one p of the world's current stock weapons.

Little Boy and Fat Man, mode the first atomic bombs dropp saki.

Intermediate nuclear forces treaty 1987

USA and USSR made the following commitment to destruction of weapons:

USSR   731 missiles   1613 warheads   445.3 megatons TOTAL
USA   350 missiles   520 warheads   94.6 megatons TOTAL

TOTAL USSR and USA commitment = 539.9 megatons
                              = 40,000 × HIROSHIMA bomb.

The fear of nuclear war is very real. Many people care very deeply about this and argue forcefully that it is wrong to devote more and more resources, including vast amounts of money, to the production and storage of weapons which could cause the destruction of our world. Others believe that the fact that nuclear weapons are so terrifying is in itself a deterrent to war. The debate on the balance between the need for the security of nations and the desire for nuclear disarmament is a highly complex one which will continue. It is a debate which should concern each of us, since the nuclear issue affects all life on planet Earth.

Demolition of Berkeley power station

Berkeley nuclear power station in Gloucestershire was opened in 1962. In 1989 it became the first nuclear power plant to face demolition. This is a complicated and highly controversial process that will take over one hundred years to complete. Ten years after Berkeley closes, the reactor will still be highly dangerous and tasks will have to be undertaken by remote control. It is estimated that it will take one hundred years for radioactivity to diminish to the level of safe doses, so that workers may enter the reactor.

Clearly this is a problem being passed on by the present to future generations.

There is much speculation over what will eventually happen to the site. Will it ever return to farm land? Will it become the site of a new reactor? Would people want to live there?

FACTS OF THE CASE:
The shutdown will generate nearly 30,000 m³ of radioactive waste that will have to be disposed of.

THE FUTURE

The whole question of nuclear power arouses intense feelings in people throughout the world, not only a fear of 'the war to end all wars' but also with regard to the nuclear industry in general. The two fundamental questions are: is it a hopeless task to try and stop or even control the development of nuclear power? And is this a desirable thing to do?

Scientist assure the world that nuclear power stations are safe, yet leaks and accidents occur often enough to undo the promises of safe, clean power. Attitudes and opinions have certainly changed with time, and possibly the incident at Chernobyl was a turning point. Successive governments throughout the 1960s and 1970s welcomed and supported the nuclear industry. In some ways nuclear power stations were symbols of a modern enlightened approach to energy problems, and so were promoted by governments.

Since that time, public opinion has had a significant effect throughout the world. In response to public opinion the governments of various countries including Austria, Sweden and Italy have decided to build no new nuclear installations. Such decisions in the United States of America are made by independent companies rather than by national government but, as a result of tremendous citizen pressure, no nuclear stations have been planned since 1979. Many people throughout the world are making it very clear that cost is not the most important consideration. Many would prefer to pay more if necessary for energy obtained from traditional sources such as coal or oil than live near to a nuclear power station.

Prospects for the development of the nuclear industry changed fairly dramatically in the 1980s, due in no small measure to the accidents in the United States of America at Three Mile Island in 1979 and at Chernobyl in 1986. When describing the Chernobyl disaster on Soviet television, Mikhail Gorbachev announced that:

It is another tolling of the bell, another grim warning that the nuclear era necessitates new political thinking and a new policy.

At the present time, the USSR remains committed to its nuclear power programme despite such grave words.

It is impossible to separate out arguments for and against nuclear power developments from those related to nuclear war. Civil and military uses of nuclear power are inevitably linked, since the reactors which generate power supplies can also be employed to produce plutonium which is used in the manufacture of weapons.

As early as 1953, particularly in the United States, there was growing public awareness of, and concern for, the fact that increased nuclear power might automatically mean an increase in nuclear weapons. The then president, Eisenhower, established the International Atomic Energy Agency which aimed to control the spread of nuclear power.

In 1968 a treaty known as NPT (the Nuclear Non-Proliferation Treaty) was signed by the USA and the Soviet Union. This treaty suggested that the world's developed nations – such as the USA, the UK and the USSR – might aid less developed countries in their attempts to establish nuclear power, provided they agreed not to use it for military purposes. Clearly this suggestion was not well received by governments who were keen to develop nuclear weapons.

Following on from the rather unsuccessful NPT, in the 1970s the United States adopted a policy of discouraging the exportation of nuclear reprocessing plants. This was intended to help the notion of non-proliferation, since reprocessing plants are the means whereby plutonium[1] – which can then be used in nuclear weapons – is extracted from used fuel. In 1976 the USA reduced reprocessing to the level necessary only for planned weapons manufacture.

Despite strong arguments against such a policy, certain countries are still pressing ahead with the development of nuclear power. France, for example, continues to expand its nuclear industry and generates two thirds of its electricity requirements in this way.

The present United Kingdom government also remains strongly committed to nuclear power. Existing power stations will be maintained and new ones are planned, for example at Sizewell in Suffolk. Sizewell was the subject of a long and bitter public enquiry, resulting in a decision to go ahead in the national interest, despite risks locally to health and environment. Older power stations are constantly monitored: one Magnox station in Gloucestershire, for example, has already closed because safety levels there did not reach the required new standards. Indeed, the cost of modernizing all Magnox stations is too prohibitive, and there is a planned programme to shut them all down before the year 2002.

In the report of the World Commission on Environment and Development, *Our Common Future*, (1987) it was concluded that the generation of electricity from nuclear energy

is only justifiable if there are solid solutions to the unsolved problems to which it gives rise.

The Commission recommended that the highest priority should be given to research and development on viable alternatives to nuclear energy and also to means of increasing its safety. Few could doubt the wisdom of this.

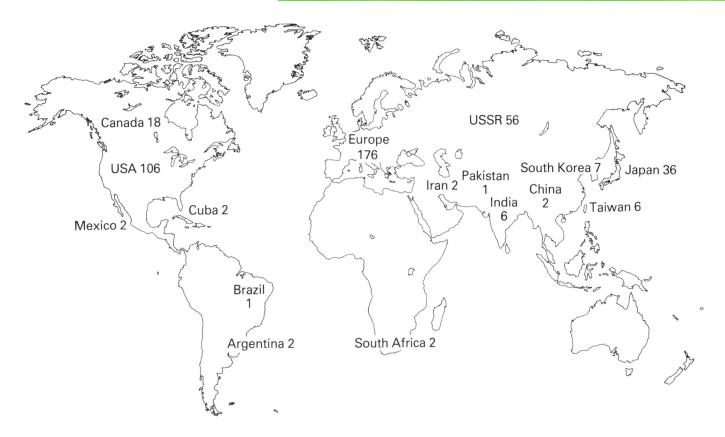

Canada 18

USA 106

Mexico 2

Cuba 2

Brazil
1

Argentina 2

Europe
176

Iran 2

South Africa 2

USSR 56

Pakistan
1

India
6

South Korea 7

China
2

Taiwan 6

Japan 36

Approximate numbers of nuclear weapons held by individual countries.

The nuclear debate will inevitably continue, with strong and sound arguments on either side.

Will future generations be safe or sorry?

Will the environment be better or worse off?

Many questions remain unanswered – hopefully this book will have helped each individual reader to clarify an informed opinion.

GLOSSARY

Atom
Constituents of all matter in the form of minute particles.

Control rods
Rods containing material that absorbs neutrons. Used to control fission.

Electron
A very tiny particle that has a negative electric charge.

Element
Elements make up all known matter.

Fission
The splitting apart of an atomic nucleus into two or more parts when a neutron strikes the nucleus. This splitting releases energy.

Fusion
Process in which two atoms fuse together.

Molecule
A group of atoms linked together.

Neutron
A very tiny particle inside the nucleus of an atom. It has no electric charge.

Nucleus
The central part of every atom.

Proton
A very tiny particle inside the nucleus of an atom. It has a positive electric charge.

Radiation
Energy or streams of particles given off by radioactive materials as they decay.

Radioactivity
The decay of an unstable atom. During this process ionizing radiation is given off.

Radon
Gas given off as the radioactive element radium disintegrates.

Reactor
A device in which a nuclear fission reaction takes place.

Uranium
An element that is used in the production of nuclear energy.

ADDRESSES — UK

British Nuclear Fuels plc
Information Services
Risley
Warrington
Cheshire
WA3 6AS

Central Electricity Generating Board
Department of Information and Public Affairs
Sudbury House
15 Newgate Street
London
EC1A 7AU

Friends of the Earth Trust Ltd
26–28 Underwood Street
London
N1 7JQ

Science Museum Education Service
National Museum of Science and Industry
London
SW7 2DD

United Kingdom Atomic Energy Authority
11 Charles II Street
London
SW1Y 4QP

UK National Radiological Protection Board
Information Services
Chilton
Didcot
Oxon.
OX11 0RQ

United Kingdom Nirex Ltd
Curie Avenue
Harwell
Didcot
Oxon.
OX11 0RH

Addresses – USA

Alliance to Save Energy
1725 K Street, NW/914
Washington, DC 20006–1401

Environmental and Energy Study
Institute
122 C Street, NW/700
Washington, DC 20001

Environmental Defence Fund
1616 P Street NW/150
Washington, DC 20036

Friends of the Earth
218 D Street SE
Washington, DC 20003

Renewable Fuels Association
201 Massachusetts Avenue NW/C–4
Washington, DC 20002

US Department of Energy
Energy Information Administration
Forrestal Building
Washington, DC 20585

US Environmental Protection Agency
401 M Street, SW
Washington, DC 20460

World Resources Institute
1709 New York Avenue, NW/700
Washington, DC 20006

Worldwatch Institute
1776 Massachusetts Avenue NW
Washington, DC 20036

Index